At the Water's Edge

AT THE WATER'S EDGE

THE SECRET LIFE
OF A LAKE AND STREAM

STEPHEN DALTON
WITH JILL BAILEY

CENTURY

London   Melbourne   Auckland   Johannesburg

Frontispiece: Grey heron (*Ardea cinerea*)

First published 1989 by Century Hutchinson Ltd,
Brookmount House, 62–65 Chandos Place,
Covent Garden, London WC2N 4NW

Century Hutchinson Australia Pty Ltd,
89–91 Albion Street, Surry Hills, Sydney, NSW 2010,
Australia

Century Hutchinson New Zealand Ltd,
PO Box 40–086, Glenfield, Auckland 10,
New Zealand

Century Hutchinson South Africa Pty Ltd,
PO Box 337, Bergvlei 2012,
South Africa

A John Calmann and King book

British Library Cataloguing in Publication Data

Dalton, Stephen
    At the water's edge: the secret life of a
    lake and stream.
    1. English photography, 1950–. Special
    subjects: Nature – Collections from
    individual artists
    I. Title
    779'.3'0924

ISBN 0 7126 2982 3

This book was designed and produced by
JOHN CALMANN AND KING LTD, LONDON

Stephen Dalton would like to thank
Wakehurst Place (part of the Royal Botanic
Gardens, Kew) for allowing him to take
photographs at the Loder Valley Reserve.

Designer Richard Foenander
Typeset by Fakenham Photosetting Ltd
Printed in Singapore by Toppan Ltd

For my children,
Joanna, Philip, Lee and Lucy

CONTENTS

INTRODUCTION

Twelve thousand years ago, the great ice sheets loosened their hold on the world's northern lands and retreated further and further north, and in their place a new, glistening world of lakes and pools, trickling streams and gleaming rivers was born. Slowly the land turned green, and as it became more fertile, new wildlife arrived. Insects thrived in the pools and lakes, and birds flew north from the warmer south to feed on them. Other animals travelled across the emerging land surface to graze on the new pastures, and following them came the predators – wolves, foxes, bears – and humans.

The world we see today is shaped by water. Valleys are carved out of the hills, cliffs and gorges cut into the rocks, as the landscape smoothed by the ice is dissected into smaller hills and river courses, each a familiar landmark to local inhabitants. People have created their own water worlds – canals and reservoirs, marinas and pools. Every settlement, even when far inland, has its village pond, today more often a tourist attraction than a focus of village life.

Despite their familiarity, there is a mysterious side to lakes, ponds and rivers. Beneath their dark surface is a secret world unseen by human eye. At times the water is like a living being, breathing columns of bubbles that burst through the surface film, sending concentric rings of ripples creeping towards the shore. It gurgles between stones, splashes over rocks, and laps gently at the bank when rippled by the wind.

It is the water's edge that stops the human tread. Here is the boundary between three worlds, where land and sky are reflected on a mirror that conceals the life below. The world of the water's edge has long held a fascination for humans. Folklore abounds with tales linking water and land, traditionally populating the former with many mysterious beings. Nowadays man seeks out lakes, streams and rivers, to sit on their banks and gaze over the water, watching the changing patterns of the reflected clouds. The water's edge remains a special place, the water itself seems accessible, yet unfathomable.

Many plants and animals make the transition from one world to the other. Kingfishers dive through the surface to spear their prey, herons stalk in the shallows, and swallows skim the water to drink. Frogs and toads, dragonflies and mosquitoes use it as a nursery, returning there to lay their eggs. Reeds and bulrushes root in the shallows, their creeping

stems edging further and further into the water, trapping mud and plant debris, building out the banks.

The habitat of the water's edge is a treasure trove for the naturalist. The bank is a haven for wetland wildflowers, and home to many small animals that hunt in or near the water – voles, shrews, ducks, warblers, and smaller hunters such as dragonflies and damselflies. Only a step away are aquatic plants, and the water is full of insects, worms, molluscs, fish, tadpoles and other water creatures, many too small to see.

Amid the lush vegetation other treasures are to be found – a dragonfly resting on a yellow flag, its wings shining in the early summer sunlight; a well-camouflaged frog lying in wait for a passing mosquito; a water vole quietly eating a willow leaf under an arching bramble. Hidden inside the stems of reeds and rushes, the larvae of waterside moths and beetles feed on undisturbed. Through these spongy tubes, air diffuses to the oxygen-starved roots anchored in the waterlogged mud. The pale nymphs of dragonflies and mayflies crawl up the stems into the sunlight, there to complete their transformation into adults.

Beyond the reeds, spiky clubrushes and the jointed stems of horse-tails invade the water. Narrow-tipped leaves of arrowhead and water plantain rise stiffly into the sunshine, and pondweeds spread broad strap-shaped leaves over the water surface, their knobbly flower heads raised high on upright stems. Large water-lily leaves form hunting perches for frogs and fishing spiders, and stepping stones for moorhens and young coots. Their undersurfaces provide sheltered sites for the eggs of pond snails and caddisflies, and often bear a rich growth of slimy algae, food for fish, tadpoles and snails.

While the water lilies are firmly anchored by creeping stems and roots, other plants float free at the surface, shifting with every current. The tiny duckweeds, their fronds only a few millimetres across, may cover large expanses of open water in calm weather. Like the water-lily leaves, they contain spongy air-filled tissues which help them to float. Tiny, thread-like roots dangle in the water, absorbing nutrients in solution. These plants are so small that they can flourish even in the water-filled footprints of cattle. The smooth carpet of duckweed on a summer pond may be deceptive; beneath its leaves may lie dark, deep water.

Below the light-stealing floating plants grow the true pondweeds which will remain totally submerged throughout their life cycle. Curly leaves of Canadian pondweed and delicate feathered fronds of hornwort and water milfoil sway in the water currents, sending streams of silvery oxygen bubbles swirling up to the surface.

The open water, shimmering in the sunlight, or stippled with falling rain, may appear devoid of life, but this is the powerhouse of the aquatic world. Tiny green and golden algae, far too small to see, grow and multiply in their millions in the sunlit surface waters. Like the green plants of the banks and shallows, the algae absorb the energy of sunlight and use it to convert water and dissolved carbon dioxide gas into living material by the process of photosynthesis. They are food for microscopic animals, often the young of shrimps, snails and fish. These, in turn, are eaten by larger animals – insect larvae, water beetles and bigger fish – which are then eaten by even larger fish, herons, kingfishers, water shrews, and so on.

In this complex web of eating and being eaten there are some strange relationships. Tadpoles, the young of frogs and toads, may fall prey to the fiercely carnivorous larvae of dragonflies. When these larvae later become adult dragonflies, they in turn may be eaten by frogs and toads, survivors of their larval hunting days.

The plants provide shade and shelter for the weaker inhabitants of the water. Few of the smaller animals venture into the open, where fish and other predators can easily see them. Among the weeds, food is to be found – algae, insect larvae and particles of dead plant material. Some animals live in the mud at the bottom of the lake or stream, feeding on rotting organic material, or lying in wait for other mud-dwellers.

As the sun filters through this underwater forest, columns of tiny oxygen bubbles stream up to the surface, produced as the leaves photosynthesize. Some oxygen dissolves in the water, a life-giving gas to be breathed in by water animals and used to burn up food and provide energy for swimming, hunting and mating, and all the bustling activity of an underwater existence.

Some pond animals steal oxygen from the air above. Pond snails float to the surface from time to time and, with a faint 'pop', open a tiny hole to let air into an internal lung. Diving beetles upend themselves at the surface and take in air to store under their wing-cases, swimming away like silvery bubbles. Water boatmen and water spiders, shrews and water voles trap a shining sheath of air under the hairs on their bodies. The water spider spins an air-conditioned underwater home, a silken bell filled with air bubbles.

On still water, the surface film is a mass of life. Surface tension, the cohesive forces between the water molecules, creates very special conditions in the boundary between air and water. Some insects spend most of their lives on the surface, almost as if on dry land. Pond skaters, water

measurers and water crickets use pads of bristles on their feet to spread their weight and walk on the water film. Whirligig beetles zoom around in crazy circles, like demented dodgem cars, rowing with their two back pairs of legs, using their front legs to seize prey. Fringes of stiff hairs turn their legs into broad paddles as they push against the water. On the back stroke, the beetles flatten the hairs, just as human rowers 'feather' their oars. At the first hint of danger – the shadow of a heron or the first large raindrops of a summer storm – they dive below the surface. At home in both worlds, they have 'amphibious' eyes: each eye is in two parts, one for seeing above the water, the other for underwater vision.

Viewed from the water's edge, the surface is unpredictable, continually changing from one mood to another as sun follows shade, or rain follows wind. It may be a reflection of the scene around it, clouds and blue sky, swaying reeds and waving branches, or a dark wind-rippled mask concealing the depths below. An idyllic water-colour scene is transformed in minutes into a ripple-smudged surrealist painting. The gentle lapping against mossy banks changes to a mayhem of foaming water, bruising reeds and rushes, tearing at undercut banks of toppling turf, and hurling pebbles against each other. It is hardly surprising, then, that the world of the water's edge attracts so many people. There are millions of tiny water worlds waiting to be explored, small oases of nature in surroundings of rapidly advancing concrete and tarmac. In even the smallest of these pools, close-knit communities of plants and animals flourish, outposts of a shrinking green kingdom, a wilder world which becomes more sought-after with its increasing rarity.

For the water world is threatened by an invisible enemy. Chemicals from fertilizers and pesticides, factories and rubbish tips, seep unseen through the ground to poison ponds and rivers, and extinguish the fragile web of life within them. Other poisons waft on the wind, dissolving in rain and water to create a sterile acid bath. The insatiable demands of the world's ever-increasing human population for homes and food crops lead to drainage of wetlands and the elimination of dewponds and natural pools, as man seeks to tame every last piece of fertile wilderness. Although he seeks out the water's edge for peace and meditation he is also destroying the water world he treasures, until one day there may be no escape for him, nor any oases in the concrete desert.

This book explores one of the surviving water worlds, a small lake created as a reservoir twelve years ago in the Loder Valley Reserve in Sussex, a gem of clear sparkling water set among some of the oldest oak woodland in the country.

SPRING

The lake lies mirror-calm, a shining sky-blue pool across which white clouds drift lazily in the early morning sunshine. A tracery of branches – alder, hazel and birch – encircles the still water, studded with new green buds unfolding in the spring air. Spears of iris leaves pierce the gleaming surface, as yet unblemished by the floating leaves of summer water weeds. The air is filled with birdsong: a chaffinch is singing in the hazel bushes, blackbirds and robins are proclaiming their territories in the wood beyond, and bluetits call excitedly as they flit from tree to tree, searching for insects in the bursting buds. The singsong voice of a lone chiffchaff among the willows signals the arrival of the first spring visitors.

A great crested grebe glides past, its handsome shades of russet, black and white blending into the rippled green water under an overhanging field maple. Tucked between its wings, snug among the warm feathers, a tiny chick raises its head to look around. Soon it will reach the safety of its home, an untidy pile of sticks and water weeds anchored to a clump of reeds.

From a perch high above the lake, the heron surveys the scene. The wood is a misty patchwork of shades of green and yellow. Dark conifers mingle with tender green hazel and alder, yellowing buds of oak, and the silver and gold of the pussy willows fringing the water. The reeds are sending up sharp young shoots from their creeping underwater stems, like an army advancing into the clear, still water. Bright clumps of marsh marigolds stud the banks, a promise of the summer sun to come. The heron has little time to stand about. Behind him, in a huge heap of sticks lined with grass and bracken, his rapidly-growing young are begging for food. He launches himself effortlessly into the air and glides down towards the water.

Beneath the gleaming reflections of sky and clouds, the lake is awakening. In its warming surface waters, a million tiny algae are soaking up the sunlight – drifting diatoms with glassy silica shells, delicately sculpted in intricate patterns; spinning colonies of *Volvox*, propelled by tiny beating hairs, whirling between thin green chains of *Spirogyra*; and microscopic flagellates flailing tiny whips behind them, twisting and turning in a never-ending dance, drawn to the sunlit surface like moths to a candle. Multiplying rapidly, they will soon be food for other creatures.

Wherever there is unpolluted freshwater there are dragonflies and damselflies. These jewel-like insects appear during the first warm days of spring, fluttering around the lush vegetation at the water's edge. Here the **common ischnura damselfly** (*Ischnura elegans*) takes a moment's rest on an iris leaf.
Dragonflies and damselflies feed as both nymphs and adults on any creature they can capture.

In the mud below, eggs are hatching and resting larvae are stirring. Newly-hatched water fleas swim jerkily up towards the surface, rowing with their antennae. The water spider leaves its silken winter cell to spin a new home among the pondweeds. Damselfly nymphs, now full-grown, drag themselves up the reed stems in the dim light of dawn. Their skins split, and they emerge slowly into the sunlight, their wings opaque and stiff. Gradually the blood pushes into their veins and their wings expand and clear, until they glint with each tremor of the breeze. Soon they are ready for their maiden flight, darting sticks of cobalt blue skimming over the water and hovering among the unfolding flowers on the bank.

Out on the lake, the water stirs and the mirror illusion is broken as a large trout leaps up in a sparkle of splashing water to catch one of the year's first mayflies, dancing uncertainly in the sun. As the sun climbs higher in the sky, oxygen bubbles stream faster from the pondweed, pricking the surface film. In the shallows, tiny specks skim across the water – pond skaters, water measurers and whirligig beetles are out hunting. Tiny rafts of mosquito eggs bob gently up and down, waiting to release the wriggling larvae into the water below.

At night, as the moon traces a silver path across the dark water, hundreds of frogs are here, some in pairs, some still searching for a mate. The water's edge is a mass of tumbling bodies, rolling in the water, scrambling on each other's backs, pushing and shoving in the frantic struggle to gain a stake in the next generation. Croaks echo through the still night air, drawing still more contestants to the lake. On succeeding nights they will be here again, the unsuccessful and the late-comers, adding their spawn to the frothy masses of jelly already bobbing in the shallows. In the deeper water, ribbons of toad spawn wind across the mud like strings of beads decorating the water weeds.

Soon, the sun-dappled shallow water is teeming with wriggling tadpoles, a welcome sight for the dragonfly nymphs and the water shrew. The banks are golden with buttercups, and cuckoo flowers and forget-me-nots are blooming behind the reeds. The water vole slips quietly out of its burrow to its favourite feeding ground, a patch of new grass nibbled to a close green sward, but vanishes again when a family of Canada geese leaves the water to bask in the sun. The sturdy yellow goslings march out of the water in single file and settle down to rest on the vole's lawn.

On a moss-covered branch – a casualty of last autumn's storms that now tips awkwardly into the water – a young coot peers uncertainly at its reflection in the water below. It is a comical youngster with huge feet, its

Several small streams meander through the surrounding woods on their way to the open water. As this part of the wood was coppiced a few years back, the floor is covered with **bluebells** (*Hyacinthoides non-scriptus*) and other woodland flowers, while the stream banks are rich with ferns, mosses and liverworts, providing food and shelter for many insects.

head covered in bedraggled plumage. The moorhen chicks are not so adventurous, but shelter inside a floating weed-lined nest.

The sound of rushing water can be heard in the distance as the stream, swollen by the spring rain, tumbles through the oakwood on its way to the lake. It snakes a twisting path between the bluebells, in places carving a deep gully between overhanging ivy-draped banks. On roots and stones a velvety coat of moss glows a fresh viridian as new shoots push their way through. Already a patchy green canopy is spreading overhead as the leaves unfold, but broad shafts of sunlight still reach to the woodland floor, adding sparkles to the water as it cascades over miniature waterfalls and bubbles against piles of loose twigs which are a legacy from the flooding waters of winter.

A tangle of twisted roots reaches down towards the water from the bank above, anchored in the cliff where the grey wagtail has made its nest. Concealed among the ivy and ferns, the young huddle together, mouths agape, waiting for their parents to return with food. Among the dead oak leaves of last autumn, the mallard has her nest. Eleven creamy eggs lie cradled in down plucked from the duck's breast, carefully camouflaged with leaves before she leaves the nest to search for food. The lake is not far away, and the young will be ready to take to the water a couple of days after hatching.

Spring passes imperceptibly into summer, and the brown banks become tinged with green as seedlings push up through the leaf litter on the woodland floor. The leaves of hazel and oak are unfurling to form a green tunnel around the stream. As the drum roll of a woodpecker sounds in the distance, the kingfisher dashes past, a streak of iridescent blue, on his way to the lake to fish. He will bring back a minnow or a stickleback as a ritual present for his mate. The birdsong is richer now: blackcaps, whitethroats and willow warblers have arrived from the far south, and are noisily setting up territories in the willows and the hazel coppice on the lake shore. Mayflies and damselflies, midges and mosquitoes form an irresistible feast, replenished every day as the larvae of the water world climb through the surface film at the water's edge, to emerge as elegant adults with shining wings.

Before the sun warms up, dew-laden **larch cones** (*Larix* sp.) glisten in the early morning light. Larches, of which there are several species, are deciduous conifers, bearing their leaves in bright green whorls.

After months of snow, ice, driving rain and even hurricane, the arrival of spring seems miraculous. Without fuss, life gently unfolds – leafy buds open, flowers, bees and butterflies appear, bird song breaks the winter silence, and fresh spring scents linger in the still, clear air.

Nowhere is spring more ravishing than around water. Here, a bubbling stream flows through a bluebell-decked Sussex valley. Below, a tuft of grass breaks through the stream bed.

The cheerful song of the **chaffinch** (*Fringilla coelebs*) is one of the most familiar bird calls to be heard in spring. Although chaffinches may be found wherever there are trees and bushes, their numbers have declined since the early 1960s owing to pesticides and herbicides. With less intensive agriculture this trend may now reverse.

It may be due to Mr Toad from *The Wind in the Willows* that toads are widely popular. The **common toad** (*Bufo bufo*) can be distinguished from frogs by its flatter back, dull, earthy colour and more warty skin. It is so well camouflaged that it may easily be mistaken for a clod of earth.

During the mating season toads converge from a wide area to their favourite breeding grounds, ignoring all other apparently suitable ponds on the way and returning to the same pond year after year. The female winds her strings of spawn around the stems of water-plants and the tadpoles hatch about two weeks later.

After the breeding season, toads wander away from the water to places where an abundance of insects can be found. Their food, wrote one Victorian naturalist, 'seems to consist of all living things that are susceptible of being swallowed'.

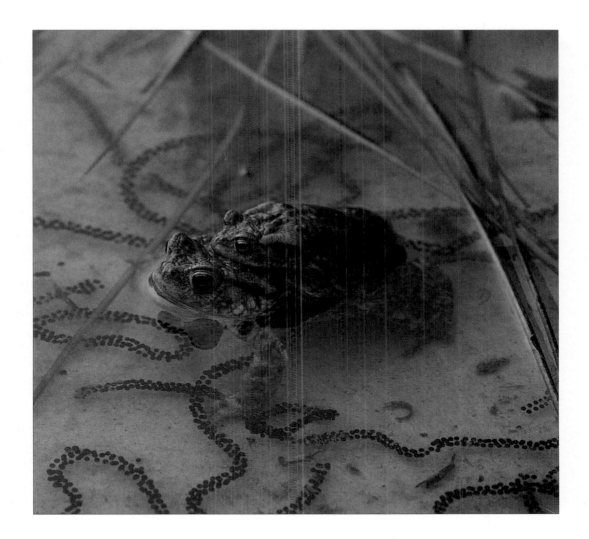

Frogs, together with most other amphibians, return to ponds and lakes to breed each spring. The **common frog** (*Rana temporaria*) lays over a thousand eggs at the bottom of a pond or ditch, but as the gelatinous covering absorbs water they swell, lighten and float to the surface.

About four weeks later the little black eggs have developed into dark brown tadpoles. After emerging, they cling to the remains of the jelly by means of a pair of suckers on the underside of the head. As they grow, gill plumes form and a mouth appears, enabling them to crop soft vegetable matter. In older tadpoles, the gill-plumes are covered over.

Ultimately the limbs develop, the hind pair first, then the front, and the gills are replaced by lungs, allowing the froglets to leave the water and live on land.

The **great crested grebe** (*Podiceps cristatus*) is a welcome and familiar sight on the lakes and reservoirs of Europe. With legs located far back under its tail, the bird is ungainly and rarely seen on land, but under water it reigns supreme, moving at surprising speed, pursuing fish to a depth of four metres and often remaining submerged for over half a minute.

The nest is a floating raft of vegetation anchored to aquatic plants or fallen branches. Young grebes are often carried on the backs of their parents to protect them from their enemy the pike.

During the nineteenth century the great crested grebe was almost exterminated as its fur-like breast feathers were used to make muffs for Victorian ladies. Fortunately the bird has become much more common since, but its future is now threatened by the increasing use of lakes for water sports.

Numerous small streams and ditches meander through the oakwood on their way to the reservoir. In the more open areas, the stream banks are decked with a variety of water-loving plants. In the foreground here are **kingcups** or marsh marigolds (*Caltha palustris*) whose brilliant glossy yellow flowers are an inch or more across. They are common and widespread around wet and marshy places in both open and more wooded habitats.

Although **mute swans** (*Cygnus olor*) are much admired for their beauty and even enjoy some royal protection, these majestic birds are extremely aggressive, frequently driving off and sometimes killing other water birds which dare to enter their territory.

Like the young of the grebe, the fluffy cygnets are regularly carried on their parents' backs to protect them from the dangers that lurk beneath the surface.

In spite of its name the mute swan is not silent, but is capable of producing a variety of calls, snorts and hisses.

A sign that spring is giving way to summer: fallen hawthorn petals drift on the water.

Most often found in damp, shady places, a cluster of **large bittercress** (*Cardamine amara*) has taken root on a moss-covered rotten log on the bank of a woodland stream.

The **grey heron** (*Ardea cinerea*) is an early nester: the eggs are sometimes laid as early as February and the young leave the nest about three months later.

As well as hunting fish, herons feed on amphibians, insects, small mammals and even birds. They often stand motionless in slow-moving water, neck stretched forward in their pursuit of prey.

The frontispiece shows a bird being buzzed by a swarm of green oak rollers – small moths, seen as tiny specks in the picture.

It is difficult to confuse crane-flies or daddy-long-legs with any other insect, with their characteristically thin bodies, narrow wings and long fragile legs that dangle beneath them when they take to the air.

Our largest species, the **giant crane-fly** (*Tipula maxima*) which can be found throughout spring and summer, has a greater wing-span than any other British fly. Its larva is semi-aquatic, frequenting the saturated margins around lakes and streams, where it feeds on plants.

Many species live around meadows and pastureland, the larvae eating the roots of cereals and grasses. They have tough skins and are commonly called leatherjackets. The adult on the right (*Tipula* sp.), was resting among the buttercups close to the water's edge.

During spring the verdant growth of ferns, **brambles** (*Rubus fruticosus*) and other plants intensifies along the stream banks. Also seen here around the water's edge are the fallen petals of **hawthorn** (*Crataegus monogyna*).

   This part of the stream supports a wealth of insect life which provides food for birds such as warblers and wrens – the kingfisher, too, can frequently be glimpsed as it flashes past skimming low over the water.

During late spring and summer huge shoals of fish fry can be seen swimming in the shallows close to the water's edge. These provide an ever-ready source of food for predators such as pike, herons and large aquatic insects.

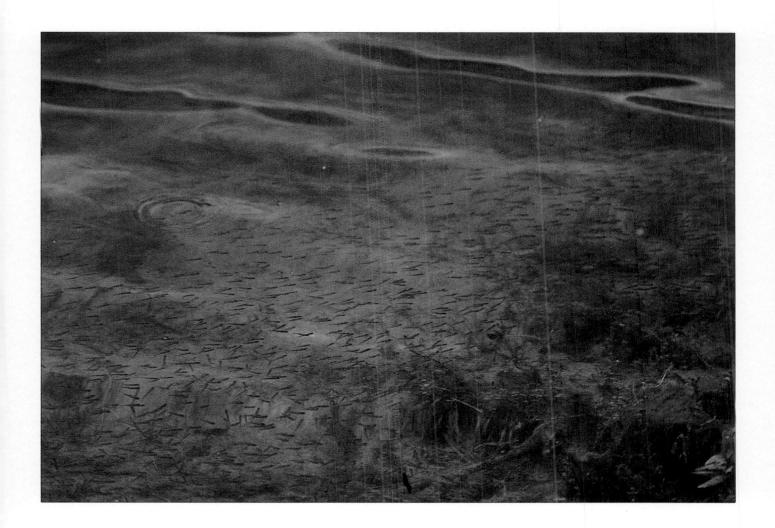

The thin film on the surface of the water is a mini-habitat of its own, supporting a host of lightweight creatures. For instance, the **common mosquito** (*Culex pipiens*) lays her spindle-shaped eggs in batches that adhere to each other, forming little rafts which float on the water. After a few days the eggs hatch into larvae which hang from the surface film by a respiratory tube at the tip of the abdomen.

There are several dozen species of mosquitoes or gnats in Europe. They all have highly complex mouthparts adapted for piercing and sucking. The males feed on nectar from flowers, while the females pierce the skin of animals to suck blood. The common mosquito feeds on the blood of frogs and birds, not man.

The **water measurer** (*Hydrometra stagnorum*) is another surface dweller and is difficult to spot, owing to its very slender body and its tendency to lurk amidst the vegetation at the water's edge. It is a true bug, with piercing mouthparts, and feeds by walking slowly on the water and detecting insect prey through vibrations. It then spears its victim with its long, thin proboscis and sucks its juices.

Embellishing the banks of streams with patches of yellow during March, the **lesser celandine** (*Ranunculus ficaria*) is one of the heralds of spring. It is a common plant, found almost anywhere which is suitably damp or shady.

Wherever there is freshwater, **moorhens** (*Gallinula chloropus*) will probably be close by. Lakes, ditches, reservoirs, streams, reed beds and farm ponds all provide suitable homes for this familiar water bird.

Although by nature shy birds, moorhens can become very tame in parks and public gardens. In their wild habitat they spend much of their time skulking among the waterside vegetation searching for water plants, berries, insects, snails, fish and the eggs of other birds. During the breeding season several broods may be reared, and the young of earlier broods may help to feed the later ones.

Newts spend most of their time on land, but are usually within striking distance of a pond or lake in which they live during the breeding season.

Our most common newt is the **smooth newt** (*Triturus vulgaris*), *right.* Outside the breeding season it is olive green in colour, but in spring the male becomes much more brightly coloured and develops a wavy crest. After an elaborate courtship dance, the female lays about three hundred eggs, and each one is individually wrapped in an underwater leaf to protect it from predators. The tadpoles which hatch some three weeks later develop in much the same way as those of the frog.

Although common in Europe, the **palmate newt** (*Triturus helveticus*), *left,* was not recognized in England until the middle of the nineteenth century. It is similar to the smooth newt in general appearance, but is smaller, its throat is not spotted and the hind feet are distinctly webbed – hence its name. Also the tail has a black filament growing out of its tip. The palmate prefers hillier country to the lowlands.

A graceful plant, **lady's smock** (*Cardamine pratensis*) is one of the first flowers to bloom in moist places. It appears around the time of the cuckoo's arrival and hence is sometimes called the cuckoo flower.

The bright, yellow-green-leaved mats of **golden saxifrage** (*Chrysosplenium oppositifolium*) are a vivid sight when the plant first appears in March. It is widely found in wet shady places, especially around springs and woodland streams.

Horsetails belong to a distinctive family of primitive leafless and flowerless perennials with tubular, jointed stems. They also produce spores on long, egg-shaped terminal cones – here the spores can be seen falling from the **common horsetail** (*Equisetum arvense*).

There can be few places more restorative to the spirit than a tranquil waterside landscape during spring. Here, water birds float lazily by and **hazel** (*Corylus avellana*) and **alder** (*Alnus glutinosa*) leaves unfurl, *left*, to the soporific hum of bees and hoverflies feeding in the trees above.

The richness of life around a lakeside is greatly enhanced if the water is sheltered, without wavelets and away from the violence of the wind. Here, the surrounding woodland provides protection.

The **mallard** (*Anas platyrhynchos*) can be found throughout the northern hemisphere and now in Australia and New Zealand, where it was introduced. Like several other species, the mallard finds its food near the water's surface and is known as a 'dabbling' duck. It has a varied diet, feeding on both vegetable and animal matter.

Contrasting with the bright plumage of the mallard drake, the duck is cryptically camouflaged with mottled feathers and a dark eye-stripe. Thus, as the picture shows, she can be extremely difficult to spot on her nest. The nest site is very variable, often on the ground amongst thick vegetation, and not infrequently in a hollow tree or a tree stump. It is made of leaves, grasses and down plucked from the duck's own breast – the drake plays no part in nest building or the incubation of the eggs and rearing of the chicks. She lays about ten elliptical greenish eggs, and when absent from the nest she covers them with down unless suddenly disturbed, when she has no time to do so.

There are several species of violet but the **common dog violet** (*Viola riviniana*) is much the most widespread, being found in open grassy places as well as woods. The banks of little streams and ditches are often literally scattered with them. Closely related to pansies, violets are perennial plants.

Close to the stream, soft young leaves of **field maple** (*Acer campestre*) have just uncurled. This is the only native maple found in England. Although it can grow to a height of forty feet (twelve metres), it is most frequently seen as a bush or as part of a hedgerow, crowded by other plants. In autumn the leaves turn a rich yellow and the tree stands out prominently from a distance.

Throughout the spring and summer months **damselflies** (sub-order *Zygoptera*) hover low over ponds and streams in their search for food or mates among the vegetation at the water's edge. Damselflies are carnivorous, though they seldom catch prey on the wing, preferring to pick up midges, gnats and other small insects which settle nearby.

Below, a damselfly is expanding and drying its wings, having just emerged from its nymphal case after spending a year or two underwater.

Seldom seen far from water, the **grey wagtail** (*Motacilla cinerea*) spends the spring and summer months around fast-flowing streams and rivers, often near waterfalls, where it may be seen flitting from rock to rock in its pursuit of insects. During the winter it moves to lowland areas, sometimes even appearing on the coast.

Grey wagtails nest on ledges or in holes close to water; the ivy-covered rock crevice here is a typical site.

The main stream flowing out of the reservoir is stony bottomed and quite fast flowing – unusual for Sussex. Numerous birds live and breed around it, including kingfishers and grey wagtails. Lining the stream's bank are alders, with exposed twisted roots reaching down into the water.

The sight of the nuptial dance of a swarm of **mayflies** (order Ephemeroptera) as it rises and falls in the warm air is a sure sign that spring is merging into summer. Clinging to the vegetation around still and running water, mayflies are amongst the most familiar of insects – their soft bodies and triangular membranous wings criss-crossed with a profusion of veins, together with the very long cerci at the tip of the abdomen, make them easy to identify.

Mayfly larvae live underwater, and take between a few weeks and three years to mature, depending on species. On emerging, the winged insect extricates itself from the nymphal case and within a few seconds flies away. The mayfly has one unique characteristic – unlike any other insect, it has two winged stages in its life cycle. Within a few hours of its emergence, the insect undergoes another moult, casting a delicate skin from the whole of its body, including wings, to reappear as a mature adult.

Spiders always feed well close to water. Here an **orb web spinner** (*Tetragnatha* sp.) has been making the most of a mayfly hatch (*left above*).

The banks of streams are invariably good places for flowers and other plants. Here, a brook is meandering through an area of once-coppiced hazelnut, with **bluebells** (*Hyacinthus nonscriptus*) and **wood garlic** (*Allium ursinum*), also known as ramsons, growing in profusion. Both these plants are members of the lily family. The stream bottom is still littered with sticks and branches – debris from winter storms.

Below, a tree root grows out of the mossy bank.

SUMMER

The water surface ripples in the warm summer breeze, transforming the reflected forms of clouds and trees, ducks and swaying reeds into a shifting medley of gleaming colours and shapes. Trailing brambles bob gently with the water, their clusters of pale pink flowers attracting bees to a feast of nectar. From the drooping tassels of the pond sedge, clouds of pollen billow into the air currents.

Since early spring, a ragged cover has been creeping over the water surface: the round, plate-like foliage of waterlilies, the narrow leaves of pondweed and pale green mosaics of duckweed and, in stagnant water, tangled mats of blanketweed. As the morning sun strengthens, the waterlily flowers open, spreading their petals to release the sweet scent of the nectar stored in the heart of each flower. As dusk approaches they close and sink deeper in the water, protected from the cool night air.

On the small stretches of clear water pond skaters and water measurers delicately tread the surface film in search of prey, and whirli-gig beetles spin in frenzied circles, spreading rings of ripples that gently rock the lily leaves. As soon as the wind ruffles the surface or a summer shower sends raindrops splattering among them, they vanish, the ska-ters and measurers leaping for the bank, the beetles diving out of danger.

Beneath the dappled reflections, the life of the lake is in full swing. The male sticklebacks are in courtship dress, with bright red breasts and bulging blue eyes. The gleaming females are swollen with eggs and the males dance before them, zig-zagging between the swaying pondweeds, sending oxygen bubbles streaming sunwards. Silver-coated water beetles weave in and out among the stems, hunting for smaller crea-tures. Snails of all shapes and sizes, round and flat, turreted and spiral, browse lazily among the weeds and lay their slimy masses of eggs under the lily leaves. Caddisflies drag their carefully constructed cases behind them, the materials varying from finely chopped leaves, twigs and weed to mosaics of tiny pebbles. But danger lurks in the shadows. Dragonfly larvae lie in wait for passing sticklebacks, and water beetle larvae sink their sharp curved jaws into the bodies of water fleas and other creatures unfortunate enough to swim within reach. Water stick insects hang vertically among the weeds, relying on camouflage for their hunting success. In deeper water, pike lie in ambush, their striped bodies

The sight of this vivid bird flashing downstream low over the water, uttering its shrill call, is an experience that never ceases to excite. The **kingfisher** (*Alcedo atthis*) is approaching its nest hole in the bank of a stream. The hole is usually excavated in a perpendicular bank, and the shiny white eggs are laid in an enlarged cavity at the far end.

blending with the dappled sunlight filtering through the water weeds. For the pike, this is the season of plenty. The sticklebacks, at the height of their mating season, are easy prey. The fry of larger fish are also growing fast, and on the surface there are ducklings and the young of coots and moorhens.

There are also visitors from the world above. The kingfisher plunges into the water in a plume of spray and the heron stalks at a more leisurely pace in the shallows, its sharp eyes alert to the tiniest movement in the water or among the reeds – a small fish, a frog, or even a water shrew. Ducks and ducklings dabble among the weeds using their large webbed feet to paddle deeper, rowing a jerky course towards the mud. The water shrew swims in a more frantic fashion, its tiny feet flailing to keep it underwater despite its life-jacket of silver air, its flexible snout twitching as it investigates weeds, stones and mud.

For some of the inhabitants of the water world, a dramatic change is taking place: they are about to transfer themselves to the world above. Young frogs, still showing the remains of stubby tails, climb cautiously onto the lily pads, still startled by the buzz of a passing fly on which they will soon learn to feed. Dragonfly nymphs, having completed their growth after the long winter, will climb the reed stems at dawn and emerge as winged adults, their bodies gleaming with colour, their eyes huge and shining, to hunt in the air above as they once hunted in the waters below. The nymphs of damselflies, mayflies and alderflies are all undergoing the same transformation. Mosquitoes are hatching out of their pupal cases at the water surface, spreading their newly-freed legs carefully on the surface film to lift their bodies free of the water as their glinting wings dry and harden.

Soon, the cycle will start all over again. The dragonflies and damselflies will search for mates, the males seizing the females behind the head with their claspers, and carrying them off. So fierce is the competition that a dragonfly may need to keep hold of his mate until she has laid her eggs, to prevent another male supplanting him and mating with her. Some pairs simply fly over the water, dropping eggs at random. Others make a laborious journey back to the world they have just left. Held firmly by the male, the female slowly climbs down a stalk into the water until she is completely submerged, to lay her eggs on underwater stems. For mayflies, the dance is the main part of the ritual. Huge swarms of newly emerged mayflies and midges dance at sunset, the buzzing of their wings attracting more insects to the group. These are all males. The moment a female appears, she is grabbed and whirled away.

Damselflies frequently rest between their brief flights – here a male **common blue damselfly** (*Enallagma cyathigerum*) is spending a few moments relaxing on a **spotted orchid** (*Dactylorhiza fuchsii*). Europe has numerous species of damselflies, many of which are predominantly blue in colour. They are smaller and more delicately built than true dragonflies, and have a feebler, more fluttering flight. They usually rest with their wings held together over their backs.

All around the lake, insect life buzzes and flutters. Pairs of shiny green mint beetles wander over the yellow flags as they mate. Butterflies sip nectar from the flowers at the water's edge and dragonflies and damselflies dart among them. Swallows skim the surface either to take the mayflies or to drink. Flycatchers hunt from perches in the hazel and alder at the water's edge, while blackcaps and other warblers hunt in the shelter of the trees. They have growing young to feed, and it is the prolific insect life of these surroundings which has brought them all the way from Africa to rear their families in the English summer. Following them came the cuckoo, its voice quieter as it prepares to return south, its fat, greedy young safely installed with reed warbler foster parents.

The banks and shallows are at their most colourful. Among the yellow flags a mist of blue forget-me-nots covers the ground, studded with the occasional pyramid of southern marsh orchid. The knobbly heads of bur-reeds and graceful panicles of reed flowers sway over the water. Wasps whine among the velvet figwort flowers. Willowherbs stand tall and pink near the bank, and the heady scent of meadowsweet pervades the surrounding woodland.

The stream no longer wanders between bluebell-lined banks, but other flowers – wood sorrel, speedwell and water dropwort-colour its edges, and tender young ferns arch over the gleaming water which flows darker now, for the sun seldom penetrates the green canopy above. Where a beam of light catches the water as it gurgles over the stones, a column of midges dances. Bees and hoverflies buzz around the brambles and alderflies crawl over the hanging leaves, laying their eggs where the hatching young will fall into the water.

Behind the tangle of roots and ivy that screens the crumbling banks, the young kingfishers huddle in their smelly nest, lined with fish bones and excrement. Deeper in the wood, the young grey wagtails clamour for food. The mallard has gone: her young are out on the lake fending for themselves.

As evening approaches, swarms of midges drift like smoke from the tops of trees and bushes, and bats emerge to feast on them. The muntjak deer leads her fawn to the water's edge to drink, their hooves leaving pairs of tiny pointed prints in the soft mud. The harsh call of a mallard echoes across the still water as the trout, which have sheltered in deeper water during the heat of the day, rise to the surface to feed. As the sun sets behind the full outlines of the summer trees, the water vole drops quietly into the water and heads for home before the owl sets out to hunt.

There are many species of sedge, but the **drooping** or **pendulous sedge** (*Carex pendula*) is the most distinctive, with its stout, tufted three-angled stems, which can grow as tall as five foot (1.5 metres), and its glossy green leaves.

The plant is widespread, found locally in damp woods on clay soils, particularly around woodland ditches and streams. This picture shows the flower-spikes shedding pollen.

As the favourite foods of the **grass snake** (*Natrix natrix*) are frogs, newts and fish, it is rarely found very far from water. It is an excellent swimmer and can remain underwater for long periods without surfacing. Grass snakes are also good climbers, occasionally even climbing trees in their search for birds' eggs.

The grass snake can grow very large, females sometimes reaching a length of four, five or even six feet (1.5–2 metres). It can easily be distinguished from the adder by its more graceful, tapering shape and the yellow collar on the back of its neck. Until the beginning of the century it was called the ringed snake – a far more descriptive name.

Flying all the way from Africa, the **willow warbler** (*Phylloscopus trochilus*) rears its young in Europe. This tiny summer visitor is a woodland bird, hunting for insects among the leaves of oak, birch and hazel, the commonest trees surrounding the reservoir.

Like most warblers, it is far more often heard than seen, with a sweet descending cadence to its song.

There is scarcely a marsh, bog, pond or stream where one or more of the numerous species of rushes does not flourish. The **common spike-rush** (*Eleocharis palustris*) is widespread and abundant in wet places. Its little flower, which is wind-pollinated, is carried on the end of a spike.

Marsh orchids (*Dactylorchis* sp.) are very variable and hybridize readily. Flowering in June and July, they are locally plentiful around marshes and wet meadows, but only a small patch thrives close to the edge of the reservoir.

The delicate, mauve-veined white flowers and pale green leaves of **wood sorrel** (*Oxalis acetosella*) make this one of the most ornamental plants of the woodland. The flowers are borne on leafless stems, while the leaves, on shorter stalks, fold up at night. The plant decorates the banks of woodland streams and other shady areas during spring and early summer.

Fringed with lush vegetation, lowland valley reservoirs, particularly those of southern England, are often rich in wildlife. After a few years sedges, rushes and reedmace extend from the shallow water up the banks to mix with the flowers and ferns of dry land. If left uncut, alders and other trees will soon appear, eventually producing woodland margins. All this plant life provides food and shelter for a host of insects, birds and mammals.

A **mute swan** (*Cygnus olor*) glides by the reed- and yellow flag-margined bank, followed by her offspring.

Although the photographs in this book all appear to have been taken in perfect conditions of sunshine, dreamy haze and no wind, in reality such idyllic conditions are becoming increasingly rare.

As a direct result of human activities, the world is experiencing dramatic changes, principally affecting the climate. The greenhouse effect is beginning to bite – our planet is heating up. Already European weather is becoming windier and more extreme.

This photograph is symbolic of the deeply disturbing changes which are likely within the next few decades affecting all life on earth.

Alderflies (*Sialis lutaria*) are frequently found resting in large numbers on waterside foliage or flying rather weakly in late spring and early summer.

The eggs are laid in clusters on plants at the water's edge and on hatching, the larvae drop into the water where they feed voraciously on small creatures on the muddy bottom. When fully grown, the larvae leave the water and pupate in the soil nearby. The pupae are insulated by spines from contact with the walls of their chamber.

On warm summer days swarms of **damselflies** (Zygoptera, a sub-order of dragonflies) may be seen flying close to the water's surface in their courtship dances, chasing each other first one way, then the other as they compete for mates.

The mating of dragonflies is unique among insects. It begins with the male grasping the female behind the head with his strong anal claspers. Mated pairs flying along like this are said to be in tandem. In many cases a brief copulation takes place on the wing, but in other instances, especially among damselflies, the pairs settle first and then form what is called a copulation wheel. Many dragonflies part as soon as pairing is completed, but in some species the male continues to hold the female by the neck while she lays her eggs.

Numerous kinds of algae play an essential part in the ecology of freshwater, providing food and oxygen for the pond inhabitants. Some types, like *Spirogyra* and blanket weed, build up into dense masses, while others are microscopic and may swim or roll around in the water. The variety and number of algae species present varies greatly depending upon the season, the chemical composition of the water and its temperature, and light conditions.

As evening approaches, shadows from **reed-mace** (*Typha latifolia*) fall across the surface **blanket weed** (*Cladophora* sp.) *below*. The bubbles of oxygen which form on the surface of blanket weed are evidence of photosynthesis taking place beneath, *bottom right*.

On long, thread-like stalks, the rare flowers of **Canadian pondweed** (*Elodea canadensis*) manage to penetrate the thick mat (*upper right*).

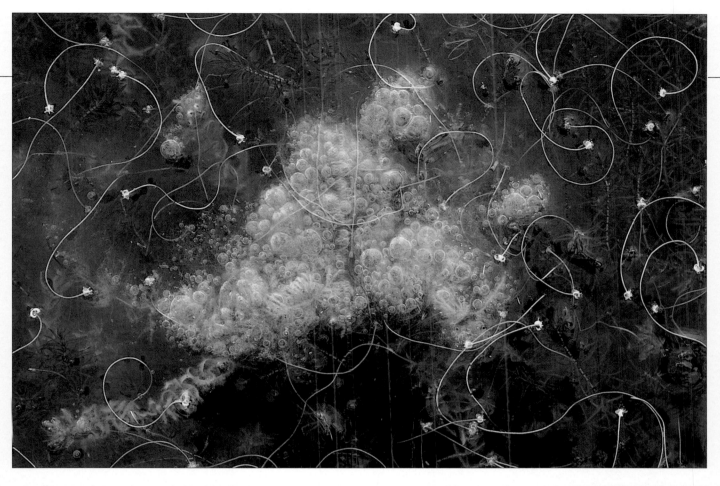

Although at first glance the **coot** (*Fulica atra*) appears similar to the moorhen, it leads a somewhat different life. The moorhen searches for food around waterside vegetation, but the coot hunts in open water, often diving below the surface to find weed and underwater creatures. The moorhen's feet are designed for climbing, while the coot's toes are partly webbed, and are more suitable for an aquatic existence. Coots are extremely territorial, and are particularly aggressive both to other birds and to their own kind during the breeding season.

The young leave the nest three or four days after hatching and are fed by the parents in the water. They do not become independent for eight weeks.

Waterside flowers always seem to be buzzing with insect life. *Left*, a **hoverfly** (*Volucella bombylans*) is feeding from a **water forget-me-not** (*Myosotis scorpioides*). There are several hundred kinds of hoverfly in Europe, and this rotund and hairy species mimics the bumble bee. In fact, there are two forms which resemble different bees, and to confuse the issue further, they interbreed. The larvae of *Volucella* scavenge in the nests of bumble bees – a good reason for the mimicry of the adult.

There is no mistaking the brilliant flower of the **yellow flag** (*Iris pseudacorus*). The plant is commonly seen in Britain and Europe around the edges of water during summer, with its ranks of sword-shaped leaves standing up to 40 inches (1 metre) high.

The strongly veined dark patches at the base of the sepals function as honey guides, attracting insects into the flower. There is also a convenient landing platform for heavy insects such as bumble bees, which allows them to probe deep into the centre of the flower for nectar.

A mating pair of **variable reed beetles** (*Plateumaris sericea*), *right*, clamber up an iris flower. Reed beetles are narrow insects, coloured with lovely metallic tints. They live on water plants, especially reeds, and their aquatic larvae obtain oxygen by inserting their tail spiracles into the air-spaces of the stems.

By midsummer the stream banks are laden with the lush growth of a host of plants. *Below*, **bramble** (*Rubus fruticosus*) and **water dropwort** (*Oenanthe* sp.) brush the water surface, swollen after a rainstorm.

The versatility of **brown rats** (*Rattus norvegicus*) is such that they can be found almost anywhere. They often wander away from the activities of man into hedgerows or follow streams and drains where food is plentiful. In fact, none of our other essentially terrestrial mammals has taken to water so well as the rat, which may easily be mistaken for the water vole. When examined closely, though, the brown rat can be seen to be much more tapered in shape, and it also has a longer tail.

The brown rat is completely omnivorous and will eat anything from fruit and seeds to birds' eggs, insects and fish.

Damselflies are smaller and more delicate in structure than the hawker dragonflies, and while at rest they hold their wings over their bodies rather than outspread like their larger cousins. They are also much less aggressive and are content with quite small territories. Consequently, they are often very numerous over a comparatively small area. The exotic-looking **common agrion** (*Agrion virgo*) is a large damselfly which haunts tree-flanked, fast-flowing streams. During courtship, groups may be seen flying in the dappled sunlight like tiny helicopters, the male often hovering over a resting female.

Although smaller, the **banded agrion** (*Agrion splendens*) is similar to the common agrion, but the male has iridescent bands on its two pairs of wings. It also prefers slower-moving streams, but the territories of the two species can sometimes overlap.

The **large red damselfly** (*Pyrrhosoma nymphula*) is very common and can be found almost anywhere from spring to summer where there is water. Here, a pair is mating.

The early morning mist over the water soon disperses as the sun rises. From spring to early summer, the rich song of the **blackcap** (*Sylvia atricapilla*) can be heard from the trees and shrubs which surround the water. Indeed, six species of warbler live in the woods and meadows around the reservoir, their ringing songs often vying with one another during spring and early summer. Until recently blackcaps were thought to be only summer migrants, but now they can sometimes be seen during winter in southern and western England, when they are visitors from northern Europe, and have learnt to supplement their normal diet of insects and berries with food from the birdtable.

Of the several species of insects which spend their lives on the water surface, the **pond skater** (*Gerris* sp.) is the most familiar. Its long legs enable it to glide across the surface film at speed, a pad of bristles at the tip preventing the leg from breaking through. It is a predatory bug, and moves around jerkily, picking up insects which have fallen onto the water and sucking their body contents dry. As well as using their eyes, pond skaters possess vibration detectors for locating struggling prey.

From early June the cobalt-blue flowers of the **water forget-me-not** (*Myosotis scorpioides*) can be seen brightening up the banks of lakes and streams. The little flowers deserve more than a casual glance – when examined closely, the exquisite form and colour of the yellow eye and white honey-guides can be fully admired.

After a freak storm, fallen silver birch trees bestrew the wooded streams which supply the reservoir. The banks are inundated with debris, and new nooks and crannies have appeared, providing nest sites and homes for all sorts of creatures.

The **water vole** (*Arvicola terrestris*) is found around the banks of lakes, ponds, canals and slow-moving streams, where it digs extensive burrows, with openings both above and below water level. Although they will eat fish, water snails and other small creatures, water voles are mainly vegetarian, having regular feeding places on stones, patches of mud or platforms of reeds where they sit up to eat, holding the food between their front paws.

Having spent three years in the mud under water, a **broad-bodied dragonfly** (*Libellula depressa*) emerges into an aerial world. After struggling out of its nymphal case, blood is pumped into the soft crumpled wings, which harden and take on a bright gleam.

Another particularly attractive butterfly is the **clouded yellow** (*Colias croceus*) and this can be seen flying in the flower-rich meadows adjacent to the water. The butterfly is a migrant to the British Isles from southern Europe and in some years it appears in profusion while in others there are none at all.

Here, one can be seen feeding on a thistle.

About twelve hours after emerging, the **broad-bodied libellula** (*Libellula depressa*) takes to the air. This dragonfly belongs to the Darter dragonfly family (Libellulidae), so called because of its habit of repeatedly darting out on a brief flight from a favourite sunny perch and then returning again.

The insect is on the wing from May to August, and is particularly common in Southern England, frequently breeding in small garden pools.

Butterflies are rarely directly associated with aquatic habitats, but the reservoir is surrounded by oak woodland and several species can be found close to the water's edge. Perhaps the most exciting is the velvety-dark **white admiral** (*Limenitis camilla*). It can be seen flying gracefully around clearings and glades in June and July in search of nectar from bramble blossoms. The female lays her eggs on honeysuckle, and during autumn the half-grown larvae hibernate, to resume feeding in the spring.

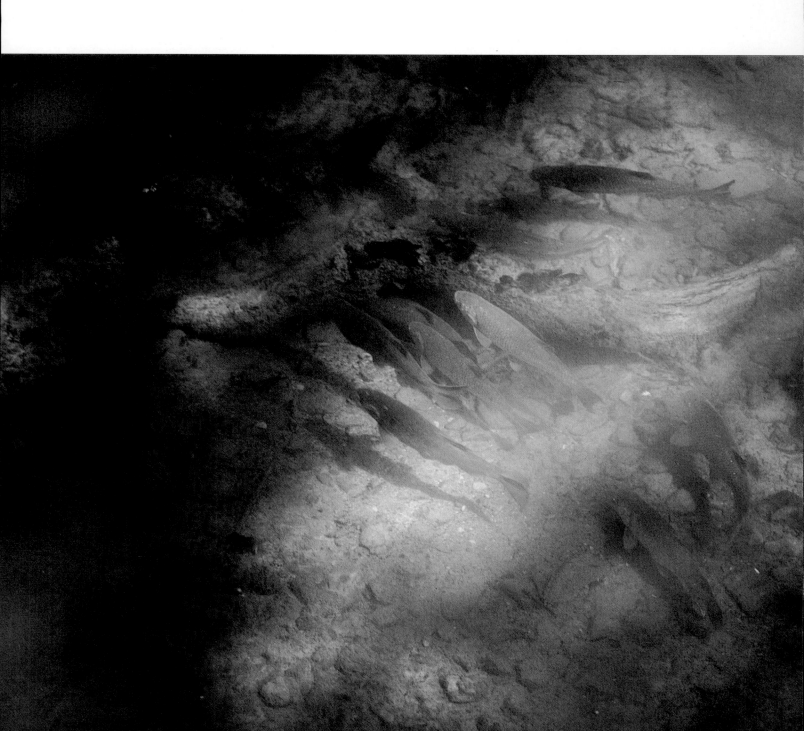

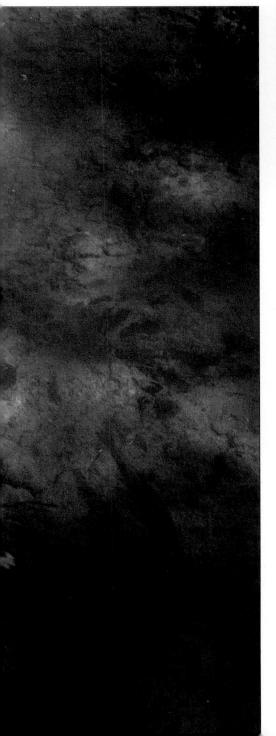

The large stream which flows from the reservoir supports several types of fish including **chub** (*Leuciscus cephalus*) and two varieties of trout (*Salmo trutta*). Here, in a pool of relatively calm water, a group of chub circle in the warm sunlight slanting between the alder trees.

Kingfishers (*Alcedo atthis*) usually hunt from a branch or other perch overhanging a stream or lake, but they are also capable of hovering before plunging into the water to catch fish. They feed on various small fish including minnows and sticklebacks, which are beaten against a branch before being swallowed head first.

Hawk moths (Family Sphingidae) are the most spectacular of all the moth families. Their thick but beautifully streamlined bodies are packed with powerful flight muscles, enabling them to fly faster than virtually all other insects. Moreover, most species are gorgeously coloured, some in bright patterns and others in subtler hues which provide camouflage when at rest. Here a **poplar hawk moth** (*Laothoe populi*) rests on a dead reed near the water's edge. This common species has two generations, the second appearing in late July.

The **giant lacewing** (*Osmylus fulvicephalus*) is an exquisite insect, with a soft, iridescent sheen on its large, flimsy wings. It is mainly nocturnal, spending most of the daylight hours resting underneath the leaves of plants bordering woodland streams. When disturbed its slow, floppy flight is quite unmistakable. The larvae live amongst the wet moss near the stream bed where they feed on the larvae of other small insects.

Around midsummer the tadpoles of the **common frog** (*Rana temporaria*) have four fully developed legs, and have absorbed their tails and lost their gills, and so need to come to the surface to breathe. They have been transformed into froglets and are ready to leave the pond. Here, one is resting on a lily pad before venturing onto dry land, where predators lurk in all corners.

Constantly changing shape as it drifts in the wind, a cloud of thousands of mating **empid flies** (*Hilara* sp.) perform in aerial courtship above the water's surface. During this captivating ritual the male lures the female with a silken parcel containing a midge or other prey. After copulation, the victim is discarded, having been sucked dry of its body juices.

Empid flies are true flies with a predatory life-style. Several hundred species are found in Europe.

Later in the summer, thistledown from the meadow wafts across the water's surface. Thistles, like many trees and flowers, rely on the wind for dispersal of their seeds.

O f all the European mammals which make their living around water, the little **water shrew** (*Neomys fodiens*) is perhaps the most enchanting. In their frantic search for insects in and around water, these creatures seem perpetually on the move, equally at home chasing whirligig beetles gyrating on the surface or rooting at the bottom for caddis larvae and worms. When underwater, the air trapped in the fur gives them a silvery appearance. To aid control and propulsion, the broad feet and toes are bordered with stiff hairs, making them into efficient paddles, while the tapering, flattened tail has a double fringe of strong hairs along the underside, functioning as both keel and rudder.

As the water shrew seldom wanders more than a few feet from the bank and never hunts in deep water, it is relatively easy to watch, if you are lucky enough to stumble across one. Water shrews are gregarious creatures and are usually found in small groups.

The **reed warbler** (*Acrocephalus scirpaceus*), like most warblers, is a migrant to northern Europe from southern and tropical Africa. Although its harsh, churring call and fast, repetitive warbling song can be heard from time to time around the reservoir, the reed warbler does not nest regularly near its shores.

In England many colonies of these birds have disappeared as a result of drainage and 'river management', which has eliminated many reed beds, their favoured nesting sites. However, the increase in numbers of flooded gravel-pits is helping to redress the balance.

Widely distributed in the northern hemisphere and found in swampy areas around ponds, lakes and streams, the **branched bur-reed** (*Sparganium erectum*) frequently forms large clumps. Each branch is made up of between two and four female flower heads near the bottom, with ten to twenty male ones above. As the fruits open, the female flower head becomes a spiky ball resembling a rolled-up hedgehog.

AUTUMN

Grey mist curls like smoke from the lake in the weak, early-morning light. Across the water the shapes of trees loom like clouds above the sentinel bulrushes. Drifting through the mist, the ghostly form of a great crested grebe glides into view, cleaving the water surface. Beneath the vapour, the water lies still and dark, the busy activity of summer slowed by the falling temperature. Even the stream is quieter, its water low after a dry August. It slips quietly into the lake between the browning stems of reeds. A robin sings tremulously among the alders.

By the water's edge the dew is heavy: golden spiders' webs are decked with pearls of water, and flowers outlined in silver while an Aeshna dragonfly rests on a reed, immobile with cold, too sluggish to shed its coat of gleaming dewdrops. Beneath the purple-pink blooms of Himalayan balsam a cranefly dangles, still after the night's dancing flights. The mallards, too, sleep on among the reeds, uniformly camouflaged in winter brown.

As a hazy sun begins to clear the mist, colour slowly returns to the water as the golds and browns of autumn softly tint its calm surface. The lily leaves are faded, sad relics of their summer beauty, with torn edges, half-eaten by snails. The duckweeds have sunk below the surface, to pass the winter as resting buds, and the stretch of open water is growing as the autumn winds tear at the lily leaves.

For some plants, the wind is a welcome ally: cottony seeds of willowherb drift along the water's edge and tangle in the spiders' webs. The bulrushes are also topped with fluffy seeds which attract flocks of migrating goldfinches, their colourful plumage glowing among the dying reeds as they feed. Tufts of thistledown tumble slowly on air currents warmed by the rising sun and float in rafts along the shore.

Other plants bear more juicy offerings. The old hawthorn, still living despite being blown down across a corner of the lake, is bright with berries. Amongst the thick ivy that drapes its slanting trunk hoverflies and wasps feed on the nectar of the ivy flowers. Wood mice climb the brambles to collect blackberries, and the dormouse gorges on the hazel nuts swelling in their ragged cups. At the edge of the lake, bumblebees hum among the balsam flowers, a last welcome feast before the winter.

At the approach of autumn, the **dormouse** (*Muscardinus avellanarius*) prepares for its long winter sleep by accumulating fat beneath its furry coat and laying up a store of nuts for the odd occasions when it may wake.

The dormouse is a surprisingly agile rodent and may be found at night scrambling around low bushes and shrubs, particularly hazel. Sadly, though, it is becoming increasingly rare.

Huge Aeshna dragonflies patrol their waterside territories, their large wings rustling. Prey is plentiful – caddisflies with papery wings, hoverflies with big garnet-red eyes and a few late speckled wood butterflies with torn wings. Delicate, pale-green lacewings shelter on the undersurfaces of the willow and alder leaves, their slender antennae waving. They will hibernate in winter in some sheltered crack and emerge in spring to produce the next generation.

Swallows swoop low over the water to catch the troops of mosquitoes which hatched at dawn. Already this summer's brood has flown south, but their parents will remain until the autumn frosts take their toll of the insects. The mosquitoes are also food for Daubenton's bat, which can be seen flitting to and fro over the water at dusk. The bats need to fatten up now in preparation for their winter hibernation.

Other animals are also preparing for winter. The dormouse and the bank vole are hoarding nuts and seeds in their burrows, and grey squirrels are busy in the wood, burying acorns among the growing pile of brown oak leaves on the woodland floor. The young water shrews born this summer are growing a thick coat of fur. They will remain awake all winter, struggling to find enough food to keep pace with their almost ceaseless activity. Their parents, having grown winter coats last year, are unable to produce such protection for a second year. They will die of cold and hunger before Christmas.

In the cooling water of the lake, other animals are also making their escape from winter. Many water insects will die before the spring, but their eggs and larvae will survive until the water warms again. The water beetle larva leaves the water and burrows into the mud at the water's brink. Here it will form a pupa and rest until spring, to emerge as a shining new beetle. The caddis larvae are spinning silk cocoons fixed to underwater stones and stems. The water spider, too, will sleep in a silken cell until spring.

On warm days, frogs and toads still come out to bask in the sun and lie in wait for passing insects. With the cooler weather, they are slower, easier prey for the heron. Soon, they will seek out warm retreats under stones where they can sleep through the winter, out of sight of birds and other enemies.

Fed by the autumn storms, the stream is once again in noisy spate, pouring over rocks and sweeping along piles of brown leaves that gradually accumulate around the stones which litter its bed. Along its mossy banks a forest of brown capsules has appeared, shedding invisible spores into the breeze. Passing flocks of chaffinches and greenfinches

As autumn drifts in, the late summer silence is broken by the rustle of falling leaves. Here, alongside the vivid greens of the moss- and liverwort-covered rocks, a withered oak leaf is carried away downstream.

splash in the shallow water, taking a quick bath before moving on to another feast of seeds.

The wet trunks of the woodland trees now stand out dark and hard as their foliage falls away. Yellow and gold leaves twist and turn as they flutter to the ground, some carried by the stream to lodge against rocky ledges and overhanging roots. They are outshone by the brilliant orange toadstools pushing up through the carpet of moss that lines the lower banks. There is moisture everywhere; slugs and snails no longer wait until the air turns cool and humid at night, but crawl out to feed by day. The atmosphere is heavy with the smells of damp wood, toadstools and rotting vegetation.

As the autumn winds usher in a cooler season, newcomers visit the lake, birds on their way south, fleeing from the cold to come. Huge flocks of noisy Canada geese land on the water with a great flurry of braking feet and wings. Water sprays over the banks, setting the spiders' webs vibrating. A water flea, held in the water droplet on a silken web, rows hopelessly as it tries to escape the tight film that traps it in an alien environment. A passing osprey pauses to fish, swooping down feet first, almost submerging as it seizes a large trout and carries it, writhing between its talons, to the safety of a tall pine. Greenshank and sand-pipers stalk the water's edge, probing the mud with their slender bills, their cries evocative of some more remote wilderness.

As the temperature falls, the first night frosts scar the remaining flowers of the water's edge and tinge the leaves with brown. While the water surface reflects the racing clouds, the world below its turbulent upper layer is falling still, the dwellers of the sunlit surface sinking into the dark mud and the water weeds becoming clogged with debris stirred up by the waves. As the wind brings a brilliant succession of gold, orange, brown and green reflections to the clear surface water, the depths dim to a still, dark underworld, a world still living, breathing, moving, but now resigned to the approaching winter, prepared for the rebirth of spring.

The quivering, slow flight of the enchanting water or **Daubenton's bat** (*Myotis daubentonii*) as it skims close to the water makes it one of the easiest bats to identify on the wing. It feeds on such insects as mayflies and caddisflies, and starts foraging about an hour before sunset.

Daubenton's bat was once considered very rare, but now appears to be relatively common and widely distributed. It may be found wherever there is a stretch of water surrounded by woodland.

As the nights become colder, morning mists gather over the water. With them come heavy dews which form myriads of glistening beads, edging every leaf, web and tendril with a silver coat. These effects are best enjoyed early in the morning before the droplets start to evaporate, and when the sun is low in the sky. At such times one realizes what enormous numbers of spiders there are around. Hardly a stem or tuft of grass is without its own little web.

Like all hawker dragonflies, the **brown aeshna** (*Aeshna grandis*) is an impressive insect, with its four-inch-long (ten-centimetre) amber-tinted wings. The insect is on the wing from late summer to October, frequently flying until sundown, often quite far from water. In England it is widespread and common, and its numbers may be augmented by migrants from Europe.

Dragonflies are sensational flyers, capable of almost any aerial manoeuvre from hovering to flying backwards. One moment they are gently floating by on rustling wings but the next instant they can accelerate and change course at breathtaking speed in their pursuit of some hapless insect.

Laden with dew after a damp, chilly night, a brown aeshna soaks up the early morning sun before taking to the air.

In the dark and damper regions, the stream banks are embellished with a wide variety of ferns, mosses, fungi, lichens and liverworts. Mosses grow especially well during the cool, moist months of autumn and winter, absorbing water over their entire surface rather than through roots and stems like conventional plants. Here, a tiny fungus penetrates a thick mat of **star moss** (*Polytrichum* sp.).

Both shy and nocturnal, the **badger** (*Meles meles*) is rarely seen except when occasionally caught in a car's headlights as it crosses the road.

Here, one is clambering down the bank to drink at the stream.

In autumn the badger starts to prepare its chamber deep underground for the cold winter months ahead by lining it with a thick layer of leaves. These gradually decompose, generating a cosy, moist warmth.

The **Indian balsam**, touch-me-not or policeman's helmet (*Impatiens glandulifera*) is a native of the Himalayas, and has become naturalized in Europe, taking seed outside the gardens where it was originally grown. Now this plant is well established by rivers and streams, where it can frequently be seen in dense masses. When ripe, the fruits – hanging pinkish-green capsules – explode violently if touched, ejecting their seeds at high speed, to be dispersed by water. The Indian balsam flowers from July right through to October and grows to a height of six feet (two metres). The flowers are a favourite with all sorts of bees. Here, a bumble bee is homing in to extract nectar.

From out of the duckweed-covered water a **common frog** (*Rana temporaria*) pokes its nose up for a breath of air. The prominent eyes with their fine golden iris and limpid black pupils are perched high up on the forehead. Unlike lizards and snakes, frogs have eyelids, but like birds they also possess a nictitating membrane, an extra, transparent eyelid. The large, brown circular patch behind and below the eye is the drum of the frog's ear. The moist, smooth skin of frogs allows respiration to carry on for an extended period without breathing through the lungs. During winter, when frogs hibernate in the mud at the bottom of the pond, all their oxygen requirements are met by breathing through the skin.

Many of the planet's life-supporting functions take place around the water surface. Micro-organisms and plants in the sea and freshwater release oxygen which diffuses to the surface and sustains animal life on land, while evaporation from the water surface is precipitated elsewhere as rain. Here, drops of rain splatter down on a **duckweed**-covered pond (*Lemna* sp.).

Frogs spend much of their time sitting on the bank of a pond or stream, but as soon as danger threatens they plunge headlong into the relative safety of the water. The enormous power of their hind legs is used chiefly for escaping predators. It only takes about a tenth of a second for a frog to extend its hind legs, propelling itself some twelve times its own length.

As these photographs show, when frogs take off the eyes are withdrawn into their sockets to protect them from damage. Here a **common frog** (*Rana temporaria*) is seen in multiflash leaping from a rock into water. Common frogs may be found anywhere where conditions are suitably moist, from open reservoirs to woodland ditches.

The **bank vole** (*Clethrionomys glareolus*) can easily be mistaken for a redder and brighter version of the field vole. However, it prefers hedgerows, dry banks and wooded country to open fields. It is also more agile than the field vole and, being an excellent swimmer and diver, it sometimes frequents wet places. By keeping still and quiet one may have the good fortune to see it rummaging around the water edges and stream banks in search of bulbs, seeds, nuts, snails and other small creatures. This one has paused for a drink before continuing its hunt for food.

Slugs (*Limax* sp.) are molluscs which have lungs but no shell. When conditions are dry they hide away in dark places, but as evening approaches, or if it rains, they come out to feed on plants and sometimes on dead animal remains. When there is no dew, slugs will often make their way to the water's edge to drink.

The tranquillity and the subtle light imbue the early morning with a crisp, magical feeling of its own. Here a **great crested grebe** (*Podiceps cristatus*) materializes from the clearing mist over the water.

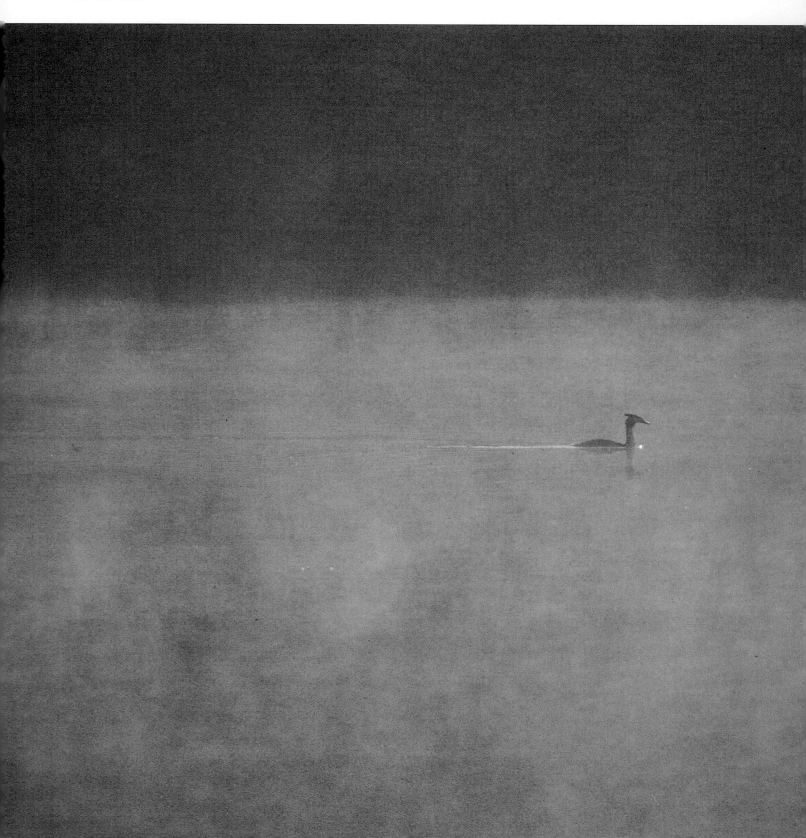

With the arrival of autumn the risk of rain and storm increases. Here the water level is beginning to rise, and soon the discarded leaves caught on the stones will be carried away downstream.

The brown rat (*Rattus norvegicus*) is an excellent swimmer and can often be found living near water, where it can become extremely skilled at catching fish. This one is seen after plunging into the swirling water

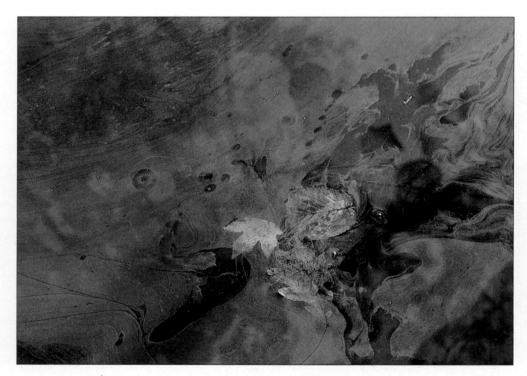

A bloom of green algae has formed on the surface of the lake, its vivid green colour mingling with the reflections of the sky and leaves above. Also three fallen leaves of oak, field maple and hazel float on the still water. Soon these will sink to the bottom and decay.

Water with lush vegetation growing around the edges is a paradise for insects. Here a **hoverfly** (*Syrphus balteatus*) feeds on the lilac-coloured flowers of **water mint** (*Mentha aquatica*). Although most abundant in summer, this attractive little fly can be seen on the wing at almost any time during the year, even in early winter. Moreover, it has been found with other insects migrating through the mountain passes of the Pyrenees and Alps.

The larvae of this and many other species of hoverfly feed on aphids, a single larva devouring as many as fifty a day.

Beads of water lie undisturbed on the leaves of reeds after a brief autumn shower.

Whirligig beetles (*Gyrinus natator*) spend much of their lives charging ebulliently around the surface of slow-moving water, diving down when alarmed. As the beetles are small, black and shiny, it is often their ripples which first attract attention. They feed on dead insects and other creatures which float on the surface. Their eyes are unusual in that they are divided into two totally separate parts, enabling them to see both above and below the water.

Before reaching the lake a narrow stream meanders through a small area of beechwood. Renowned for their autumn colour, beech leaves start yellowing in early October, changing to orange-brown before falling around early November. Here the stream banks are littered with beech and hornbeam leaves, while clumps of pendulous sedge appear here and there.

Masses of freshly fallen autumn leaves collect around the stones of the stream bed. The full splendour of these varied hues only lasts a day or two, as the colours very rapidly fade to brown and eventually almost black as the leaves become water-logged and are attacked by the unrelenting action of micro-organisms.

WINTER

Jagged blue fingers of ice steal across the water, entombing the lake dwellers in a translucent twilight. The surface dims, its reflections blurred and distorted. Around its margin, bare branches are silhouetted against the winter sky, writhing in the gusts of sleet-laden wind. Withered brown leaves from the woodland edge spiral into the air and skitter across the ice to be trapped in pools of still unfrozen water.

Deceived by the thin film of rainwater over the ice, a mallard tries to make a landing. It skids helplessly, its webbed feet slithering in all directions. Swans and Canada geese are still enjoying what remains of the open water, bobbing stoically on its choppy surface as the sleet flattens their feathers and dribbles off their bills.

The rain creates patterns in the water, a fine frosting effect in drizzle, craters and coronets in the first large raindrops of a winter storm, and a heaving turmoil of wavelets as the storm gathers force. On calm grey days it becomes a cold gleaming mirror, as the gnats dance over their own reflections in the weak winter light.

Beneath this changing surface, the life of the water world continues. Pond snails crawl slowly over the decaying pondweeds. Dragonfly and mayfly nymphs lie in wait among the weed, creeping into the warmer mud in cold weather. Sheltering in the deeper water, away from the chilly surface, the fish are sleepy: they feed little and breathe slowly, saving their energy until food becomes more plentiful.

As the ice spreads over the water, the light below becomes dimmer and the oxygen supply decreases. But the ice is welcome to the inhabitants of the lake. Above this crystalline shield, the air temperature may fall drastically, but the water below remains warmer, unfrozen. In the mud, winter buds of duckweed and creeping stems of reeds and yellow flag rest, swollen with food to fuel the spring growth. Here, too, lie caddis larvae and the eggs of water fleas and other small creatures. The water shrew is at home under the ice, carrying its air supply with it as it dives. If food becomes scarce, it will supplement its diet with slugs and earthworms. Safe under a large boulder well above flood level, the toad sleeps away the winter, still and cold.

On calm misty days, the monochrome landscape takes on a beauty of its own. The fine stems of the reedbed resemble the shading of a

A reed stands stoically amid the thickening ice. The translucent sheet shields the underwater plants and animals from the colder air. Without this insulating layer many of them would perish.

lithograph, interspersed with the dark heads of bulrushes. In the background, pine trees create shadows among the tracery of winter branches. Like a highlighting pencil, the hoar frost paints the margins of leaves and twigs, revealing details hitherto unnoticed: the twisting patterns of the dried grasses, the clusters of black ivy berries set among the leaves. As the sun gently dispels the mist, a million tiny crystals glint and sparkle, each like a perfect faceted gemstone, glassy clear.

Slowly the mist clears, and life stirs at the water's margin. Slugs and snails emerge to feed on the decaying vegetation. The moorhen leaves the shelter of the reeds, picking her way carefully over the ice to the open water. The coots are out in force, having migrated to southern England from the colder continent. In a quiet corner of the reservoir, where the incoming stream has kept the water ice-free, the herons are fishing, spaced out along the bank like anglers.

As the sun strengthens in the new year the coots become restless, and squabbles break out as they try to establish territories in preparation for the turbulent days of courtship. Furious chases are continued far out on the water as the coots, heads lowered, lunge at each other with angry squawks. The heronry is also noisy: courtship is at its height. The mallards are in spring plumage, the glossy emerald necks of the drakes iridescent in the sunlight. There is much head-bobbing and bill-dipping as they pursue the females across the water.

Swollen by the winter rain, the stream is a welcome source of sound and movement in the stillness of the winter woodland, cascading between the dark bare trees. It eddies under the thin casing of ice in the shallows, and gushes over miniature waterfalls, the trapped air bubbles gleaming white below the flood, to burst into a spray of sparkling droplets over a large boulder. The brown rat is also out scavenging, scampering sure-footed across a thin root that overhangs the stream. The air is heavy with the woody smell of decay. Toadstools push up through the moss, and form frills along fallen tree trunks that bridge its banks.

Out on the lake, the water remains bare and dark: no lily leaves or duckweed enliven its surface, no fish rise to send ripple rings rolling towards the water's edge. It is like those empty waters into which the poisons of the human world have seeped, creeping through the drainage ditches, or raining from the polluted skies above. But this lake, surrounded by the nature reserve, is no dying body of water. Here, the watcher on the winter bank can still gaze over its apparently lifeless surface and dream of summer days when mayflies will skim the surface and the trout leap in response, and bees will buzz among the flowers.

As water splatters onto a boulder, the cold drops slice through the clear winter air.

Against a background of firs, a **mallard** (*Anas platyrhynchos*) takes off from the water to find an alternative feeding ground.

Ducks are not only amongst the fastest of flying birds, but also capable of long migratory flights.

During early winter before the water becomes too cold, aquatic insects still may be quite active – here a **mayfly nymph** (order Ephemeroptera) lurks among blanket weed, where it feeds on plant debris and algae. The feathery gills on the nymph's abdomen beat continuously, wafting a current of water over the body – oxygen is also thought to be taken in through the cuticle, or outer shell.

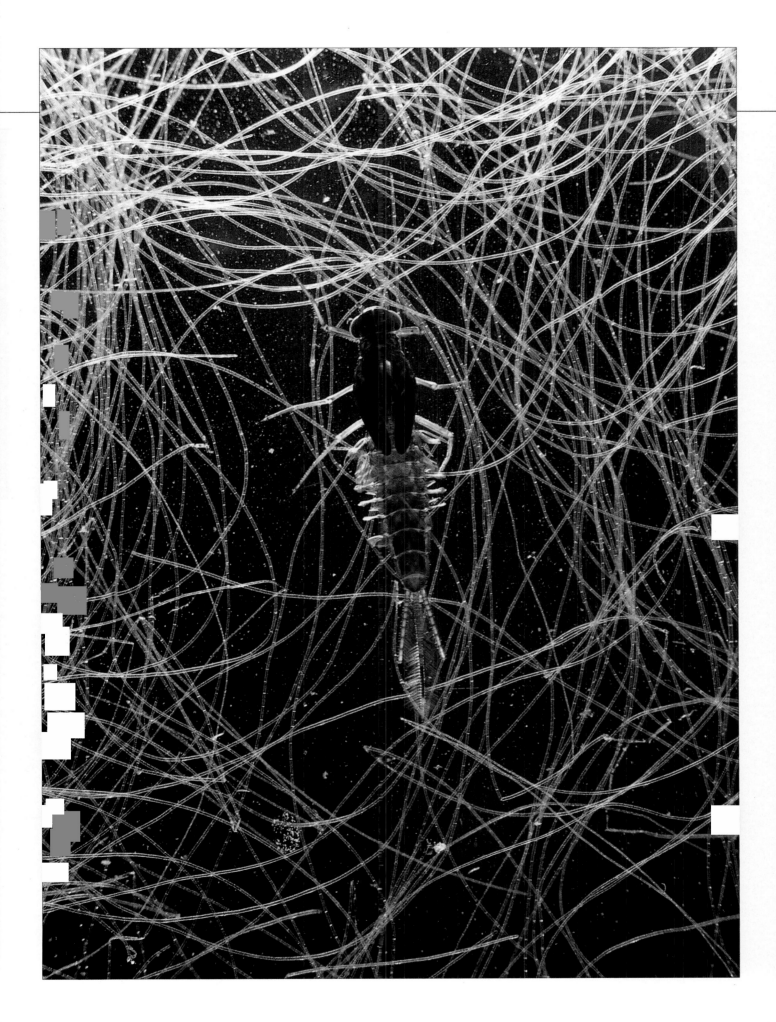

Some nooks and crannies of the stream bank are oozing with water rising from the small springs and ditches draining from the woody hills above. Here, a dangling tree root encrusted with saturated moss drips relentlessly into a pool of still water.

A tangle of tree roots trail in the water, exposed by the scouring of winter floods. The banks of Sussex clay readily disintegrate in the swirling floodwater.

Unless the winter is very cold many underwater creatures remain active. The **water stick-insect** (*Ranatra linearis*) is found in shallow water in Europe and southern England. The long tail is not a sting but a breathing tube, which penetrates the surface to supply air to the creature while submerged, but the insect can inflict a painful, stabbing bite with its powerful beak. A well-camouflaged predaceous water bug, it lies in wait to ambush unsuspecting tadpoles, small fish and other insects which come within reach of its 'jack-knife' front legs. If conditions in the water become unsuitable the insect crawls out, opens its wings, and flies elsewhere.

Pond snails always seem to be active, except when the water temperature is near freezing. They creep over vegetation and stones, searching for algae and other plants. Snails feed by using a file-like structure in the mouth called a radula, which scrapes off plant fragments in a steady rhythmic motion.

The **great pond snail** (*Lymnaea stagnalis*) can reach a length of two inches (five centimetres) and is common in large ponds and lakes in Britain and Europe.

Dragonfly nymphs (*Aeshna* sp.) are very aggressive predators, hunting among water plants well below the surface for other insects, fish and even small frogs. To catch their prey they have an extendable lower lip called a mask, which is shot out at lightning speed by a sort of hydraulic action, and seizes the victim in the formidable pincers at the end. This hawker nymph has just caught a stickleback.

Not much is known about the lives of dragonfly nymphs during the winter. When the water becomes cold they retreat to the deeper regions, where it is unlikely that they are very lively, but in milder conditions they are more active, feeding a little and growing slightly. The severity and duration of the winters affects the time spent as a nymph.

Bounding from stone to stone, a **wood mouse** or long-tailed field mouse (*Apodemus sylvaticus*) makes its way across a stream. It is an extremely agile rodent, moving along in a series of zig-zag bounds with short pauses, holding its long tail off the ground. It is also a capable climber, often clambering high into bushes in search of berries, nuts or even birds' eggs. During winter it is less active, having collected up quantities of beech mast, acorns and other nuts and seeds and stored them in hiding places underground, or sometimes in an old bird's nest.

The wood mouse has many enemies, but prolific breeding maintains its numbers – the females can produce up to six litters a year between March and November. It is a common mammal, occurring in woods, fields and gardens all over Britain and Europe.

The sound of water as it cascades over the stream's rocky bottom breaks the stillness of the winter day.

Picked out by crystals of hoar frost, these dead reeds display a fascinating pattern of twisted leaves. In the gentle warmth of spring, new shoots will emerge from the creeping stems buried in the mud.

With the approach of spring the days lengthen as the sun creeps higher in the sky. In the early morning light, a cob **mute swan** (*Cygnus olor*) begins his territorial display, parading his domain with partly raised wings. Already the sound of bird song from the surrounding woods can be heard drifting over the water.

ABOUT THE PHOTOGRAPHS

As this book is about a small patch of land and the life which depends on it, I have avoided explanations of the photography. Suffice to say that a wide range of approaches and techniques were adopted to tackle the variety of subjects portrayed. However, for those interested I have included a brief summary of the basic technical information relevant to each photograph.

Clearly, equipment will not produce pictures on its own – the make and complexity of the camera have little to do with the final results and more often than not, the simpler the equipment the better. All the photography could have been taken with almost any manually focused and exposure-controlled single-lens reflex.

The beauty of the natural world can only be revealed to the eye and camera by spending time in the field. Above all, it requires patience, light and an understanding of nature.

KEY

Format	35 mm Nikon F3, FE2 or Leicaflex SL	35 mm
	2 ¼″ square Hasselblad	2 ¼″
Lens	Focal length in millimetres	000 mm
Film	Kodachrome KII 25,64 or 200	KII K25, K64, K200
	Fujichrome 50	F50
	Ektachrome 64	E64
Shutter speed and aperture		⅟60 sec. f16
Special high-speed shutter	An asterisk indicates that a special high-speed shutter was used	⅟500*
Flash	Electronic flash used in combination with daylight or as sole light source	& flash
	High-speed flash (⅟10,000 sec or faster)	& H/S flash

page 37 Cranefly
35 mm 105 mm K25 ⅛ sec f8

38 Stream
35 mm 24 mm K25 ¼ sec f16

39 Fish fry
35 mm 105 mm K200 ¹⁄₁₂₅ sec f8

40 Mosquito eggs
35 mm 55 mm (reversed)
K25 ¹⁄₈₀ sec f8–11 and flash

41 Water measurer
35 mm 100 mm K25 ¹⁄₈₀ sec f16
and flash

42 Moorhen at nest
35 mm 200 mm K64 ¹⁄₁₂₅ sec f5.6

43 Lesser celandine
35 mm 105 mm K25 ½ sec f11

44 Palmate newt
35 mm 105 mm K25 ¹⁄₈₀ sec f11–
16 and flash

45 Smooth newt
35 mm 100 mm K25 ¹⁄₈₀ sec f11–
16 and flash

46 Lady's smock
35 mm 105 mm F50 ⅛ sec f8

47 Common horsetail
35 mm 100 mm K25 ¹⁄₈₀ sec f11
and H/S flash

Golden saxifrage
35 mm 100 mm K25 ¹⁄₃₀ sec f11
and flash

48 Hazel leaves
35 mm 100 mm K25 ¹⁄₃₀ sec f8–
11 and flash

49 Lake
2¼" 50 mm F50 ⅛ sec f16

50 Mallard
35 mm 400 mm K64 ½ sec f5.6

51 Nest of mallard
35 mm 24 mm K25 2 sec f16

52 Violets
35 mm 100 mm K25 ⅛ sec f8–11

53 Field maple leaves
35 mm 100 mm K25 ¹⁄₆₀ sec f8
and flash

54 Damselfly in flight
35 mm 135 mm K25 ¹⁄₅₀₀* sec f11
and H/S flash

55 Damselfly emerging
35 mm 100 mm K25 ¹⁄₆₀ sec f11
and flash

56 Grey wagtail
35 mm 135 mm KII ¹⁄₆₀ sec f16
and flash

57 River
2¼" 50 mm F50 ¼ sec f16

58 Mayfly
35 mm 105 mm K25 ¼ sec f8

Mayflies in web
35 mm 100 mm K25 ¹⁄₈₀ sec f11–
16 and flash

59 Mayfly in flight
35 mm 135 mm K25 ¹⁄₅₀₀* sec f11
and H/S flash

60 Stream
2¼" 50 mm F50 ½ sec f16

61 Root
2¼" 135 mm F50 4 sec f22

63 Kingfisher in flight
2¼" 150 mm F50 ¹⁄₅₀₀ sec f16 and
H/S flash

65 Damselfly on orchid
35 mm 100 mm K25 ¹⁄₈₀ sec f11
and flash

67 Pendulous sedge
35 mm 100 mm K25 ¹⁄₈₀ sec f11
and H/S flash

68–9 Lake
2¼" 50 mm F50 ⅛ sec f16

70 Grass snake swimming
35 mm 105 mm K25 ¹⁄₈₀ sec f11
and H/S flash

71 Willow warbler in flight
35 mm 135 mm K25 ¹⁄₁₂₅ sec f11
and H/S flash

72 Spike rush
35 mm 100 mm K25 ¹⁄₈₀ sec f8
and flash

72 Marsh orchid
35 mm 105 mm K25 ¼ sec f11

73 Wood sorrel
35 mm 105 mm K25 ¼ sec f16

74 Swan with cygnets
35 mm 400 mm K200 ¹⁄₅₀₀ sec
f5.6

75 Pond
2¼" 50 mm F50 ⅛ sec f22

76 The greenhouse effect
2¼" 50 mm F50 4 sec f22

77 Alderfly in flight
35 mm 135 mm K25 ¹⁄₅₀₀ sec f11–
16 and H/S flash

78 Damselflies courting
35 mm 400 mm K200 ¹⁄₅₀₀ sec
f5.6

79 Damselflies mating
35 mm 100 mm K25 ¹⁄₁₀₀ sec f11

80 Lake with blanket weed
35 mm 55 mm K25 ¼ sec f16

81 Canadian pondweed
35 mm 105 mm K25 ¼ sec f8

Oxygen bubbles
35 mm 105 mm K25 ⅛ sec f11

82 Coot
35 mm 200 mm K64 ¹⁄₆₀ sec f8

83 Young coot
35 mm 400 mm K200 ¹⁄₂₅₀ sec
f5.6

84 Hoverfly
35 mm 100 mm K25 ¹⁄₆₀ sec f11
and flash

Yellow flag
35 mm 105 mm K25 ¹⁄₁₅ sec f8

85 Reed beetles
35 mm 105 mm K25 ¹⁄₆₀ sec f11
and flash

86 Steam bank
35 mm 55 mm K64 ¹⁄₁₅ sec f11

87 Rat
2¼" 135 mm F50 ¹⁄₅₀₀ sec f16 and
flash

88 Banded agrion
35 mm 100 mm K25 ¹⁄₆₀ sec f16
and flash

Red damselflies mating
35 mm 100 mm K25 ¹⁄₈₀ sec f11
and flash

89 Common agrions courting
2¼" 135 mm F50 ¹⁄₅₀₀ sec* f16
and H/S flash

90/1 Lake in mist
35 mm 300 mm K64 ¹⁄₁₂₅ sec f4.5

91 Blackcap singing
35 mm 400 mm K200 ¹⁄₁₀₀₀ sec
f5.6

92 Pond Skater
35 mm 100 mm K25 ¹⁄₈₀ sec f11–
16

93 Water forget-me-not
35 mm 105 mm K25 ¹⁄₁₅ sec f11

94 Stream
2¼" 50 mm F50 ¼ sec f16

95 Water vole
35 mm 280 mm K2 ¹⁄₆₀ sec f11
and flash

96/97 Dragonfly emerging, sequence
35 mm 100 mm K25 ¹⁄₃₀ sec f11
and flash

98 Dragonfly in flight
35 mm 135 mm K25 ¹⁄₅₀₀ sec* f11
and H/S flash

page 98 White admiral
35 mm 100 mm K25 ⅟₈₀ sec f8
and flash

99 Clouded yellow
35 mm 100 mm K25 ⅟₆₀ sec f11
and flash

100 Chub
35 mm 100 mm K64 ⅟₆₀ sec f5.6

101 Kingfisher
2¼″ 150 mm F50 ⅟₅₀₀ sec f16 and
H/S flash

102 Poplar hawk
35 mm 100 mm K25 ⅟₆₀ sec f11
and flash

103 Giant lacewing
35 mm 100 mm K25 ⅟₈₀ sec f11
and flash

Young frog
35 mm 100 mm K25 ⅟₆₀ sec f11
and flash

104 Empid flies courting
35 mm 105 mm K64 ⅟₅₀₀ sec f4

105 Thistle seeds
2¼″ 135 mm F50 ⅟₁₅ sec f11

106 Water shrew
35 mm 100 mm K25 ⅟₈₀ sec f16
and flash

107 Water shrew diving
2¼″ 135 mm F50 ⅟₅₀₀ sec f16 and
H/S flash

108 Reed warbler
35 mm 135 mm K2 ⅟₁₀₀ sec f11–
16 and flash

109 Branched bur-reed
35 mm 100 mm K25 ⅟₁₅ sec f8

111 Dormouse
2¼″ 150 mm F50 ⅟₅₀₀ f 16 and
flash

113 Rocky stream with fallen leaf
35 mm 55 mm K25 ¼ sec f11

115 Daubenton's Bat
2¼″ 150 mm K64 ⅟₅₀₀ sec f11
and H/S flash

116/117 Waterfall
2¼″ 50 mm F50 3 secs f22

118 Lake with mist
2¼″ 50 mm F50 ¼ sec f16–22

119 Spider and web
2¼″ 135 mm F50 ⅟₃₀ sec f8

120 Brown aeshna with dew
35 mm 100 mm K25 ⅟₆₀ sec f8–
11 and flash

121 Brown aeshna in flight
35 mm 135 mm K25 ⅟₅₀₀ sec* f16
and H/S flash

122 Fungus and star moss
35 mm 105 mm K25 ½ sec f11

123 Badger
2¼″ 150 mm E64 ⅟₂₅₀ sec f11
and flash

124 Indian balsam
35 mm 100 mm K25 ⅟₃₀ sec f11
and flash

125 Common frog in duckweed
35 mm 105 mm K25 ⅟₈₀ sec f16
and flash

Raindrops on water surface
35 mm 105 mm K64 ⅟₁₂₅ sec f16
and flash

126/127 Leaping frog (multiflash)
2¼″ 150 mm F50 ½ sec f11 and
H/S flash

127 Frog diving into water
35 mm 100 mm K25 ⅟₈₀ sec f11–
f16 and H/S flash

128 Bank vole drinking
35 mm 100 mm K25 ⅟₈₀ sec f11
and flash

129 Slug
35 mm 100 mm K25 ⅟₈₀ sec f11
and flash

130/1 Great crested grebe
35 mm 300 mm K64 ⅟₁₂₅ sec f5.6

132 Swirling stream
2¼″ 50 mm F50 1 sec f22

133 Rat swimming
2¼″ 135 mm F50 ⅟₅₀₀ sec f16 and
H/S flash

134 Algal bloom and leaves
2¼″ 135 mm F50 ⅟₁₅ sec f11

135 Hoverfly on water mint
35 mm 100 mm K25 ⅟₁₂₅ sec f11

136 Beads of water on reed
35 mm 105 mm K25 ⅛ sec f11

137 Whirligig beetles
35 mm 100 mm K25 ⅟₆₀ sec f11
and H/S flash

138 Woodland stream
2¼″ 50 mm F50 ¼ sec f22

139 Leaves in stream
2¼″ 50 mm F50 2 sec f22

141 Ice
35 mm 55 mm K25 ⅛ sec f16

143 Water spray
2¼″ 135 mm F50 ⅟₅₀₀ sec f8–11
and flash

144/145 Lake
35 mm 24 mm K25 ½ sec f16

146 Mallard in flight
35 mm 400 mm F100 ⅟₂₅₀ sec
f5.6

147 Mayfly nymph
35 mm 100 mm K25 ⅟₈₀ sec f11–
16 and flash

148 Water drop
2¼″ 135 mm F50 ½ sec f11–16
and H/S flash

149 Tree roots
35 mm 55 mm K25 ¼ sec f11

150 Water stick-insect
35 mm 105 mm K25 ⅟₈₀ sec f11–
16 and flash

151 Pond snail
35 mm 105 mm K25 ⅟₈₀ sec f11–
16 and flash

Dragonfly nymph with fish
35 mm 100 mm K25 ⅟₈₀ sec f11–
16 and flash

152 Wood mouse
35 mm 100 mm K25 ⅟₆₀ sec f11–
16 and H/S flash

153 Wood mouse jumping
2¼″ 135 mm F50 ⅟₅₀₀ sec f16 and
H/S flash

154 Water spray
2¼″ 135 mm F50 ⅟₅₀₀ sec f8–11
and flash

155 Frost on dead reeds
35 mm 105 mm K25 ⅛ sec f11

156/157 Swan
35 mm 400 mm K64 ⅟₅₀₀ f5.6